THE FARM

CROPS ON THE FARM

Ann Larkin Hansen

ABDO & Daughters

Published by Abdo & Daughters, 4940 Viking Drive, Suite 622, Edina, Minnesota 55435.

Printed in the United States.

Cover Photo credits: Peter Arnold, Inc.
Interior Photo credits: Peter Arnold, Inc.

Edited by Bob Italia

Library of Congress Cataloging-in-Publication Data

Hansen, Ann Larkin.
Crops on the Farm / Ann Larkin Hansen
p. cm. -- (The Farm)
Includes index.
Summary: Describes the main crops grown in the United States, the soil composition and climate conditions necessary for their growth, the developement of hybrids, and the importance of caring for the land.
ISBN 1-56239-625-0
1. Crops--Juvenile literature-- 2. Agriculture----Juvenile literature. 3. Farms.--Juvenile literature. [1. Crops. 2. Agriculture. 3. Farms]
I. Title. II. Series: Hansen, Ann Larkin. Farm
SB102.H36 1996 96-12456
633'.00973--dc20 CIP
AC

About the author

Ann Larkin Hansen has a degree in history from the University of St. Thomas in St. Paul, Minnesota. She currently lives with her husband and three boys on a farm in northern Wisconsin, where they raise beef cattle, chickens, and assorted other animals.

Contents

Crops Feed the World

Driving through the countryside on a fine summer day, you see miles of green fields. Some may be corn or soybeans, or green hay bending in the breeze. You might see huge wheat fields, hundreds of apple trees, or a half-mile of tomato plants. All these growing plants are the crops that feed us and our animals.

All the food we eat comes from crops. Even our meat, eggs, and milk come from animals that were fed crops grown especially for them.

Opposite page: Rolling farmland with hay stacks.

Kinds of Crops

Most people think of crops as **grain** crops. A grain is the seed of a plant. We eat the seeds of corn, wheat, oats, barley, and rice. Fruits and vegetables are also crops. So are things like cotton, cocoa, coffee, sugar beets, and pepper.

Each type of crop needs the right kind of **soil**, sun, and water. Farmers must know what kind of crops will grow in their soil and **climate**.

Opposite page:
A farmer harvesting beets.

Kinds of Dirt

Not all **soil** is the same. Some is light-colored and sandy, and dries out quickly. Some is black and slimy, and seems to be wet at all times. The best soil for crops is in-between: not too heavy, sandy, dry, or wet. This type of dirt is called **loam**, and it grows the very best crops.

Sandy soils are good for **root crops**, where the **edible** part grows underground. Peanuts, potatoes, and carrots grow best in sandy soil. Rice grows best in wet soils. Most other plants grow well in loam.

Opposite page:
A farmer in Japan planting rice, which requires very wet soil.

Kinds of Climate

Plants can be fussy about the weather. Oranges and cotton need long, hot, wet summers. But oats and peas prefer cooler weather. Wheat doesn't care if there isn't much rain. Many types of apples grow best where the winters are cold.

The type of **soil** and the kind of **climate** on a farm tell the farmer what types of crops to grow. This is why different crops are grown in different parts of the country.

Opposite page:
In Texas where it is hot and dry, carrots thrive in the sandy soil.

Southern Crops

The **growing season** is the time between the last **hard frost** in the spring and the first hard frost in the fall. A hard frost is when the temperature falls below freezing, and plants can no longer grow.

In the southeastern states, the growing season is very long and the summers are wet and hot. Long-season crops like celery, sugarcane, rice, peanuts, oranges, grapefruit, cotton, and tomatoes grow well here.

Southeastern growing region.

Opposite page: A celery field in Florida.

SOUTH BAY
CELERY
FRESH CELERY
40-12139

The Grain Belt

Iowa, Illinois, Nebraska and nearby states have the richest **loam soils** in the country. The **growing season** is shorter than in the South, but there is plenty of rain. This is corn and soybean country. These states grow so much **grain**, they are called the "grain belt."

Most of the corn and soybeans are not grown for people. The grain is fed to hogs, **cattle**, and chickens. But some is made into corn chips and breakfast cereal.

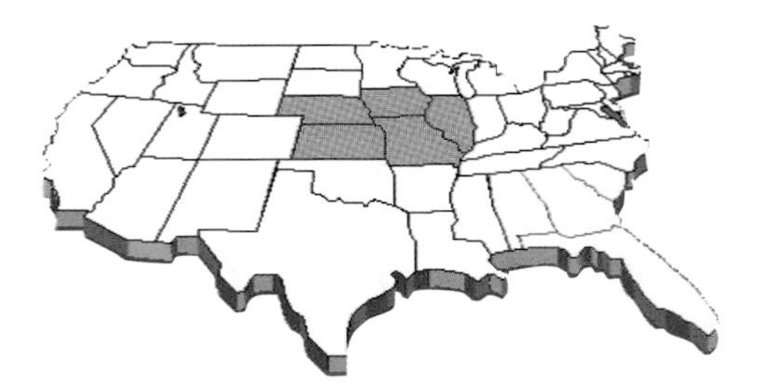

Midwestern growing region.

Opposite page: Combining corn on a farm in Iowa.

JOHN DEERE

The Northern States

States to the north and east of the **grain** belt have colder weather and hillier land. These areas are better for **dairy** cows than corn. Crops that grow in cool, rainy weather are planted here.

Potatoes and apples do well in certain areas. Large amounts of oats and hay were once grown in the northern states. But plant scientists have developed corn that grows in cooler areas. Now more farmers in this area are growing corn instead of oats.

Northern growing region.

Opposite page: The lush farmland of the North.

The Dry West

West of the Mississippi River, it is too dry for corn or vegetables. This is wheat country, and the fields are the biggest in the nation. In many areas, there is only enough moisture to grow a crop every other year. Half the land is left **fallow**, or unplanted, each year.

Because of the lack of rain, most western states grow wheat. Wyoming, Colorado, and Montana raise **cattle** instead of crops.

Western growing region.

Opposite page: A wheat field in the western growing region.

Irrigation

A farmer **irrigates** the field by pumping water from the ground or a river. Warm, dry California irrigates more **acres** than any other state. It is the biggest grower of crops that love hot weather and lots of water. California is the top producer of cotton, grapes, lettuce, and most vegetable crops.

Other dry states such as Texas and Nebraska use lots of irrigation to grow cotton and corn.

Opposite page: A field being irrigated in California.

The Green Revolution

In the 1960s, scientists worried that there would not be enough food to feed the world's population. So they developed new **hybrid** plants. Hybrid corn, rice, and other crops grew more food on the same amount of land.

The new hybrids spread around the world, causing the "Green Revolution." Today, scientists continue to work on **breeding** better plants.

Opposite page: Hybrid corn harvest.

Plant Problems

Plants can get sick and die just like people. Many **acres** of crops are lost to bad weather each year. Many more acres are damaged by poor **soil**. Good farmers know that healthy crops need healthy soils.

A healthy soil is full of earthworms and **bacteria**. There is plenty of **minerals** that help plants grow. **Chemicals** that kill crop **pests** can kill soil, too. Many farmers are using fewer chemicals.

Opposite page: Rich, healthy soil in Oregon.

Looking After the Land

There is only so much farm land in the world. If the land is well-cared for, it will always grow crops to feed us.

Good farmers know how to keep the dirt from blowing away in the wind, or washing away in the rain. They plant fields with different crops each year to keep **pests** confused. They spread **manure** and **fertilizer** to replace **nutrients** the crops have used up. Taking good care of the land means plenty of crops and good food for everyone.

Opposite page:
A tea field in Japan.

Glossary

acre—an area of land equal to 43,560 square feet (13,277 m).

bacteria—tiny organisms that can be seen only with a microscope.

breeding—the process of generating or growing new plants.

cattle—animals of the ox family, such as cows, bulls, and steers raised for meat and dairy products.

chemicals—man-made substances that are applied to crops, either to promote growth or kill pests.

climate—the pattern of temperature and weather typical of an area.

cross-breeding—crossing different types of the same plant to get better offspring.

dairy—production of milk and related products.

edible—fit to eat.

fallow—land plowed but not planted for a year or more.

fertilizer—anything spread on soil or crops to aid growth.

grain—the small, hard seeds of cereal crops (wheat, corn, oats, rice, and soybeans).

growing season—the number of days between the last hard frost in the spring and the first hard frost in the fall.

hard frost—when the temperature falls below freezing (usually overnight) long enough or far enough to kill most annual plants.

hybrid—a cross between two different types of a plant, usually producing an offspring that is stronger or more productive. Hybrids usually don't "breed true;" that is, if you plant a hybrid seed, the resulting plant will not look like its parent.

irrigation—pumping water through pipes, ditches, and channels from underground, rivers, or lakes to water crops.

loam—soil composed of about equal parts of clay and sand, usually with a good portion of organic matter (decaying plants, etc.).

manure—animal waste. Used as fertilizer in the fields.

minerals—substances that occur naturally and are neither animal or plant. Crops need large amounts of phosphorus, potassium, and nitrogen, and trace amounts of other minerals such as calcium and magnesium. Fertilizers contain minerals.

nutrients—things that nourish plants and help them grow.

pests—an insect or animal that is destructive to crops or plants.

root crops—crops that have the edible portion growing underground. Examples are potatoes, carrots, and radishes.

soil—part of the Earth's surface that plants are grown in; dirt.

Index